IMAGES
of America

SQUANTUM AND SOUTH WEYMOUTH NAVAL AIR STATIONS

When the men serving at the Squantum Naval Reserve Aviation Base looked for an appropriate image for their logo, they researched the history of the Boston area. Given Quincy's proximity to Salem, they chose the Squantum witch for patches, tail insignia, and even the base newspaper.

IMAGES of America

SQUANTUM AND SOUTH WEYMOUTH NAVAL AIR STATIONS

Donald Cann and John J. Galluzzo

Copyright © 2004 by Donald Cann and John J. Galluzzo
ISBN 978-0-7385-3624-8

Published by Arcadia Publishing
Charleston, South Carolina

Printed in the United States of America

Library of Congress Catalog Card Number: 2004104739

For all general information contact Arcadia Publishing at:
Telephone 843-853-2070
Fax 843-853-0044
E-mail sales@arcadiapublishing.com
For customer service and orders:
Toll-Free 1-888-313-2665

Visit us on the Internet at www.arcadiapublishing.com

To all the military and civilian personnel who worked to protect our country, serving from the air bases at Squantum and South Weymouth.

The men of Naval Air Station South Weymouth selected one of Massachusetts' most enduring symbols of the state's fighting spirit: the silhouette of the Lexington and Concord Minuteman.

Contents

Introduction 7

1. The Harvard Air Meet 9

2. The Victory Destroyer Plant 21

3. The Squantum Naval Air Reserve Base 35

4. Building the South Weymouth Station 55

5. South Weymouth during World War II 71

6. From V-J Day to Decommissioning 81

7. South Weymouth Airshows 107

Acknowledgments 127

Comdr. Jack Shea (far right) spent more than 10 years with the naval reserves at Squantum and eventually served as executive officer of the base. He died fighting a fire on USS *Wasp* as it sank off the Solomon Islands on September 15, 1942. His heroic actions lingered in the memories of the Squantum crews for a long time. Shea's effect was recognized even after Squantum closed in favor of South Weymouth. His legacy remained strong enough for the U.S. Navy to name the main thoroughfare into the base off Route 18 in his honor.

INTRODUCTION

In another era, the land once used as Naval Air Station South Weymouth would have been considered a ghost town. Today, no jets streak into the sky, no U.S. Navy or Marine Corps pilots train for deployment to military hotspots around the world, and no crowds gather to experience the power, speed, and elegance of American military aviation in action at airshows. A decade ago, the base was alive with human activity. Today, what was known as the noise of the daily life of a naval air station is gone.

The same can almost be said for New Squantum, the former site of the Squantum Naval Air Station in Quincy. Yet, while the South Weymouth land awaits final word of its future and the inevitable development that will help diminish the memory of the navy's presence even more, Squantum has already seen its future arrive. Gone are the biplane races of the second decade of the 20th century, the destroyers of the Victory Destroyer Plant, and the roaring engines of the PBY Catalina patrol planes. There are instead condominiums, restaurants, and the stores of Marina Bay. Very little remains that can be used to tell the stories of the men and women who worked and trained on those grounds: the Wright brothers, Amelia Earhart, Lt. (later Adm.) Richard E. Byrd, and many others.

The Squantum and South Weymouth naval air stations will be forever linked due to their geographical proximity and their intertwined operational histories. Squantum came first with a reserve training facility in the 1920s, upgraded during the Great Depression. South Weymouth then opened as a lighter-than-air facility in the 1940s but closed down briefly after World War II. Due to the growth of Boston's Logan International Airport, the navy opted to close Squantum in favor of South Weymouth in the 1950s, moving to the inland site and expanding it to fit the needs of the jet age. That site remained operational until 1997.

The world changed dramatically in the years documented in this book. In 1910, at the time of the Harvard-Boston Aero Meet, America was not yet a superpower. However, Pres. Theodore Roosevelt had, by that time, sent the White Fleet around the world as a show of the country's military potential. The flyers buzzing around the sky at the first air meet had no idea that within a decade the world would be at war and the grounds from which they had taken-off would be transformed into a wartime industrial facility, churning out destroyers ostensibly for use in World War I, but most of which saw action in later conflicts.

The most menacing tool used by the German enemy in World War I, the U-boat, dictated the development of antisubmarine warfare technology at both Squantum and, later, South Weymouth. As the United States transitioned from World War I into the Great Depression and the next world war, the specific task of hunting U-boats urged competition between developers of seaplanes, lighter-than-air blimps, and (later) helicopters to find the ultimate sub-tracking and sub-killing machines. When World War II ended, the focus shifted to the submarines of the Soviet Union and the armed conflicts of the cold war. The sons and daughters and grandchildren and great-grandchildren of the U-boat hunters of World War I flew the same aerial pathways of their forebears decades later. Only after Communism had collapsed around the world did the United States downsize its military nationwide and called for the closing of several installations, including Naval Air Station South Weymouth in 1997.

Between 1910 and 1997, America went from biplanes to blimps to long-range fighter bombers and stealth fighters; from fighting Fascist Germany to Communist Russia to Ba'athist Iraq; from isolationism and neutrality to becoming the world's premier military power.

The South Shore towns of Quincy, Weymouth, Rockland, and Abington became forever altered by the growth and expansion of both naval air stations. What was once an open field known as New Squantum was transformed from an idyllic and quiet seaside settlement to a hotspot for early aviation, a record-breaking naval production facility, the birthplace of the American naval air reserve, a concurrent civilian airport, and finally a trendy private community. At South Weymouth, an entire way of life changed forever when the Old City vanished in favor of the lighter-than-air base. Later, when the navy saw the need for expansion, a major thoroughfare between Weymouth and Rockland (Union Street) was separated in favor of a looping, circuitous route around a new runway. Highly trafficked neighborhoods instantly became secluded dead ends on either side of the base. Every time a driver passes down Veterans of Foreign Wars (VFW) Drive, he or she is reminded that somewhere a Russian submarine demanded attention, hence the separation of Union Street.

There is no possible way that the entire story of either the Squantum or South Weymouth naval air stations could be told in the format of this book. There simply is too much to tell. Two excellent histories, *Navy Wings over Boston* and *The Defender's History: A Historical Account of Naval Air Station South Weymouth, Mass.*, reach much further in depth than this book ever could. Our goal is to bring the story of both stations to the general public in an easily accessible format, and to do so with some never-before-published photographs from local private collections. We cannot promise that every unit that ever flew out of either station is mentioned in our text, nor can we promise that every plane seen in the skies over Squantum and South Weymouth will be represented. Yet, we can promise that we have done our best to give the reader an idea of what life at each station was like and to give a basic history of what existed prior to the development of each facility.

One
THE HARVARD AIR MEET

It had been more than six years since the Wright brothers proved to the world that man could fly, and the excitement generated by the opening of the first Harvard-Boston Aero Meet in September 1909 proved that aviation was not just a passing fancy of the American public. Gate receipts for the first show, which featured such famous fliers as England's Claude Grahame White (shown here taking off in his Farnam biplane) totaled approximately $150,000. Thousands flocked to the field at Quincy's New Squantum to watch the show. Aviation had arrived in New England.

Mingled in with the crowd at the air meet were several recognizable faces. Boston mayor John "Honey Fitz" Fitzgerald, grandfather of Pres. John F. Kennedy, knew that a large gathering of people meant a large gathering of voters. He made his way through the crowd with handshakes and, no doubt, campaign promises, although there is no indication that he sang his famous campaign stop song, "Sweet Adeline," while attending the air meet.

Massachusetts governor Eben Sumner Draper made the rounds as well. For Draper, a self-made man born to a manufacturing family in Hopedale (outside of Worcester), the Harvard Air Meet signaled the nearing end of his popularity in state politics. A proponent of harbor and river improvements and education reform, he alternately took a negative stance on labor unions. After two successful elections, he soundly lost in his third attempt to become governor.

Several other dignitaries and people of importance attended the 1910 meet as well, including Pres. William Howard Taft and the first family, and Charles Elliot, president of Harvard University, the sponsor of the meet. Taft also brought along his new secretary of the navy, George von L. Meyer, who had previously served as postmaster general under Pres. Theodore Roosevelt. Gen. Nelson A. Miles, born near Westminster, Massachusetts, and wounded four times during the Civil War, made an appearance as well. Rising quickly through the ranks during the war and fighting with the Army of the Potomac at many important battles such as Antietam, Fredericksburg, and Chancellorsville, Miles eventually took the position of general in chief of the army in the 1890s, leading the nation's troops in the Spanish-American War and other conflicts. The most recognizable face in the crowd is in this photograph. Just two years out of law school and a newly elected a state senator from New York, 28-year-old Franklin Delano Roosevelt looked on with great interest. His knowledge of the potential military uses of airplanes would be of great use to him during his stint as assistant secretary of the navy, starting in 1913, and his years as president of the United States during World War II.

Understanding the obvious link between aviation and military tactics, flyers like Claude Graham-White demonstrated the practical possibilities of combat from the cockpit. Here, White shows how he planned to drop a bomb, in this case flour and plaster of Paris, from his Farnam biplane onto a mock battleship. His accuracy caught the attention of von L. Meyer.

The Harvard Air Meet attracted all sorts of airmen, including Wilbur and Orville Wright. The brothers attended the meet and even flew, although they did not do so competitively, opting instead to allow stunt flyers to pilot their craft. Here, a Wright plane is brought onto the grounds in preparation for flying. When the Wrights flew in 1903, their longest flight lasted 59 seconds. At the Harvard Meet, pilot Ralph Johnstone set a new American record for remaining airborne for 3 hours, 5 minutes, and 40 seconds.

Charles Willard, of Melrose, Massachusetts, flew while Lt. Jacob Fickel practiced firing from above. A month earlier, Fickel had fired the first bullet ever shot from an airplane at a practice field near New York City.

Just four months prior to the Harvard Air Meet, aviation pioneer Glenn H. Curtiss flew from Albany to New York City. On May 29, 1910, Curtiss became the first man ever to fly between the cities, and the first to use the Hudson River as an aerial navigation tool. Known as the father of aviation for his many contributions to the field, including an early dirigible, Curtiss shared his work with men such as Alexander Graham Bell, aviation enthusiast and inventor of the telephone.

While the monoplanes, biplanes, and triplanes ruled the day, various other forms of human flight were successfully attempted. This demonstration of a human kite was given long before parasailing became a standard offering of Caribbean vacation packages. The Harvard Air Meets (others were held in 1911 and 1912) became so popular that one man, Russell K. Green, wrote a song about them, "Take Me Down to Squantum, I Want to See Them Fly": "Now there is a girl as bright as a pearl / Who's crazy about aviation / To each aeroplane show she will go with her beau / She knows every airman in creation / Each day just as morn, does o'er the bay dawn, / She is waiting outside the field gates / The-a-tres and wine are not in her line, / And this song she sings while she waits / Take me down to Squantum I want to see them fly / Billie Hoff and Beachy flying round so high / We'll eat peanuts and popcorn and watch them in the sky / We'll ride down in the car or we'll walk, it's not far / If you'll take me to Squantum Field."

Eighteen-year-old Cromwell Dixon wowed the crowd with his lighter-than-air craft. The dirigible lost power at one point and nearly sent him drifting out over the ocean. However, that was, comparatively, the least of his worries that day.

Competing for a $5,000 prize, Dixon accepted the challenge to fly from Squantum to the statehouse in Boston and back. Unfortunately, because he simply did not know his Boston landmarks well enough, he mistook the Christian Science church for the statehouse and came back only to find out he had been disqualified for veering off the route.

English pilot A. V. Roe's experimental triplane came to an unfortunate end during the show, one of only two damaging mishaps during the weeklong program of events. The pilot sustained no injuries as a result of the accident, nor did the pilot of a Pfitzner monoplane that crashed upon takeoff.

The organizers of the meet had safety provisions in place just in case something unexpected should happen. Here, in a lifeboat, apparently just out for a spell on the water are, from left to right, an unidentified coxswain, a Mr. Belcher, Claude Graham-White, Sidney MacDonald, and A. V. Roe.

Harriet Quimby, one of the future darlings of aviation, attended the first Harvard Air Meet (1910), where she supposedly gained the inspiration to take to the skies herself one day. She did just that, becoming America's first licensed female aviator and the first woman to fly across the English Channel. Tragedy struck for Quimby, though, during the Harvard Air Meet in 1912, when she attempted to beat the record set by Claude Graham-White for flying from Squantum to Boston Lighthouse and back twice in succession. She lifted off on July 1 with a passenger aboard, William Willard, one of the organizers of the first two air meets. Thousands watched in horror as first Willard and then Quimby fell out of the plane and tumbled through the air 1,000 feet to their deaths. The catastrophe marked the end of the air meets at Squantum.

Although Quimby and Willard's deaths ended the enthusiasm of the early air meets at Squantum, civilian aviators returned to the area in 1927. The federal government, which by then operated first a destroyer plant (see chapter 2) and a naval reserve airfield (see chapter 3) at Squantum, leased nine acres of land to Quincy architect Harold T. Dennison for use as an airfield.

Dennison offered flight training classes, airshows, and scenic excursions over the coastline. In October 1923, Dennison had a training accident at the new Squantum base when he ditched into the sea as his wing clipped a sandbar in Dorchester Bay. He and his training officer swam and crawled to shore, where they were taken to the hospital. His own training classes stressed safety precautions while in the air. Here, Doris Shalit-Oberg and Paul Wilcox stand outside the Dennison hangar.

Doris Shalit-Oberg became the youngest woman to fly solo out of the Dennison airfield in 1930, when she was still attending Quincy High School. Amelia Earhart took lessons at the Dennison field in 1927, just a few short months before being the first woman aviator to cross the Atlantic Ocean.

Harold Dennison put much of the work of his airfield into the trusted hands of Harold Martin. From the time of its opening on September 3, 1927, until its last flight on June 17, 1941, the Dennison airfield witnessed the training of approximately 150 pilots a year and the formation of several aero clubs, which included glider clubs and a group of Chinese American pilots.

The popularity of American airmen and their gravity-defying bravado continued unabated into the 1920s and 1930s. When news leaked of Spaniard Juan de la Cierva's new autogyro in the late 1920s, American Harold Pitcairn jumped at the opportunity to be the first American to manufacture them. He began production of three prototypes in the fall of 1929. This autogyro appeared at the Dennison field on April 1, 1931, just seven days before Amelia Earhart set a world's altitude record of 18,415 feet. Exactly two weeks later, pilot Jim Ray landed on the White House lawn and took off again while his boss, Pitcairn, received an award from Pres. Herbert Hoover for his role in the development of the aircraft type. The autogyro reached its height in popularity in 1934, when Cary Grant landed in one at his wedding in the movie *It Happened One Night*. The hybrid plane, seemingly half plane, half helicopter, signaled a transition in American aircraft design and the ever widening advances in the field of aviation. Such advances eventually (at the beginning of World War II) sounded the closing bell for Dennison's field and opened a new chapter in the life of the Squantum peninsula.

Two

THE VICTORY DESTROYER PLANT

This undated aerial view of Squantum Point was most likely taken soon after the Victory Destroyer Plant was built in 1918. The United States did not enter World War I until April 1917, but in response to one of the most dreaded weapons in the German arsenal, the U-boat, the country began to build up its naval forces as early as 1916. The navy acquired 630 acres of coastal Quincy marshland in October 1917 and contracted the Bethlehem Steel Corporation to build an emergency annex to their Fore River Shipyard on the site. The government, committing $15 million, financed the construction of the plant, which turned out to be the world's largest destroyer plant. Within seven months, the first keel was laid (April 20, 1918) and the first ship was launched (July 18, 1918) and commissioned USS *Delphy* (November 30, 1918) in Boston. The war had ended 19 days earlier. The plant built 35 Clemson-class destroyers from April 1918 until June 1920, when the government took back control of the plant. The last destroyer built was USS *Osborne* (DD-295), launched on December 29, 1919, and commissioned on May 17, 1920.

The winter of 1917–1918 was cold, and ice formed on Dorchester Bay. The weather conditions did not stop a work force of 4,600 that worked night and day to complete the huge undertaking of building the destroyer plant. The Victory Destroyer Plant was designed to build only destroyers and to build them indoors using a module method of construction. This process was new and innovative to the shipbuilding industry.

This floating crane was a workhorse in the construction of the plant. The main building had 10 covered building ways. There were also six wet slips, two 25-ton cranes, and four 5-ton cranes. Many other buildings and facilities were constructed. Other industries went to work to supply the plant. The Field Point Boiler Plant of Providence, Rhode Island, sent 90 boilers, and 35 came from the Black Rock Plant in Buffalo, New York.

Shown here are the fitting-out wharves and the main plant building. After the ships were launched, they were tied up to these wharves to be fitted out. When this was finished, they were taken to Boston to be commissioned and to receive their new crews.

This construction photograph gives a good look at conditions that prevailed during the building of the plant. Note the ice on the water and the number of men and materials. The navy even built a wooden bridge to Dorchester to make it easier for the workers to get to work each morning. The bridge lasted until 1925, when the navy took it down, causing a loss of convenience for the residents of Squantum.

In addition to filling in the swamp, the dredging of the slipways was done. Concrete pilings were poured to support the mostly steel structures built near the water.

This is the main plant building under construction. As is evident in this photograph, it was a steel-frame building. This structure eventually had thousands of windows.

This view of the floating crane gives some details of the heavy equipment used in the construction of the plant. The tugboat to the right of the barge likely worked as hard as the crane while maneuvering it around in very icy waters. The sign on the barge to the left gives an idea as to how far some of this equipment came to build this plant.

This is the main plant building seen from the inland side. The framing is mostly done, and the frames for the windows are in place. The words "Bethlehem Steel" are clearly marked on the railroad cars in the foreground.

25

With a construction work force of more than 4,000, a place to eat lunch and dinner was well appreciated. The workers are eating off trays, indicating that hot meals were available.

By June, ice in the bay was just a memory of the long winter. Many things had to be attended to from waterside, such as guiding large pieces of steel hoisted by the crane or getting to a bolt in a difficult place, hence the collection of boats.

This is a photograph of the completed main office building, built after the destroyers were started. In 1920, the plant was closed. Kennedy Marine Basin leased 516,000 square feet for a marine business in 1930. On June 22, 1932, the plant caught fire and was completely destroyed. Navy personnel from the Squantum Naval Air Station, led by Lt. John J. Shea, lost their battle against the fire. Mahlon W. Walker and his chums from the 1932 graduating class of Quincy High School came to help on the next day. Already in their white suits for the graduation exercise, they went to the fire, broke some windows to let the smoke out, and arrived at graduation on time, with their suits presentable.

With a work force of about 4,000 to construct the plant and 8,000, including 150 women, to operate the plant, and shifts that went around the clock, this on-site recreational facility made a lot of sense.

This is Squantum Street and the gate to the plant. German saboteurs may have been a threat, but the security at the gate looks a little thin.

This hull, No. 341, became USS *Delphy* (DD-261), the first ship launched from this plant. *Delphy*'s five-year service was mostly uneventful until September 8, 1923. Assigned to San Diego, it was the flagship of Destroyer Squadron 11 and, on September 8, was in the lead of seven destroyers traveling from San Francisco to San Diego. An incorrect navigation decision by *Delphy* in bad weather caused all the ships to crash upon the shore.

This is the hull of the destroyer *Delphy*, shown coming down the ways at its launching on July 18, 1918. Waiting for it are tugboats and a spectator fleet. The vessel was named after Richard Delphy, a War of 1812 midshipman killed on August 14, 1813, in the action between the *Argus* and HMS *Pelican*.

Delphy is in the water, and the tugboats have taken charge. They will guide the vessel to a dock, were it will be fitted out. Note the two men in a canoe. Launching day is a time of joy and hope for the success of the new ship, especially the first ship down the ways of this plant. All the good feelings of this day ended when *Delphy* broke in two on September 8, 1923, and left three dead on the rocks at Point Honda, California.

This is the hull of USS *Osborne* (DD-295), launched on December 29, 1919. It was named after Lt. Weedon E. Osborne, who was killed in 1918 while trying to help an injured officer. For this action he received the Congressional Medal of Honor. *Osborne* was decommissioned in 1930 and sold for scrap on January 17, 1931, in order to fulfill requirements of the London Naval Treaty of 1930. The Standard Fruit and Steamship Company later converted its hull into the banana boat *Matagalpa*. The vessel later became an army transport and caught fire in 1942. It was scuttled off Australia in 1947.

This is the hull of USS *Thornton* (DD-270), launched on March 2, 1918. It was named after James Shepard Thornton, executive officer on *Kearsage* during its engagement with the Confederate ship *Alabama* off the coast of France during the Civil War. The *Thornton* was inactive from 1922 to 1940 and was then recommissioned as a seaplane tender (AVP-11). It was at Pearl Harbor Submarine Base on December 7, 1941. It went into action and with the other seaplane tenders claimed one Japanese torpedo bomber with no casualties. The vessel was beached after colliding with other ships on Ryukyus in 1945.

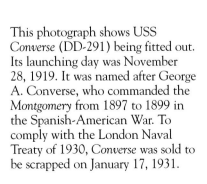

This photograph shows USS *Converse* (DD-291) being fitted out. Its launching day was November 28, 1919. It was named after George A. Converse, who commanded the *Montgomery* from 1897 to 1899 in the Spanish-American War. To comply with the London Naval Treaty of 1930, *Converse* was sold to be scrapped on January 17, 1931.

This is likely the winter of 1918–1919, and it looks as harsh as the one before. The two destroyers at the fitting-out dock are not identified. The photograph gives a good view of the above-water lines and the superstructure of a Clemson-class destroyer, known as flushdeckers or fourstackers during World War I and between war periods. This class replaced the Wickes class of destroyers. The Clemson class had a longer steaming range and could travel 2,500 nautical miles at 20 knots. These ships displaced 1,190 tons of water and were 314 feet long by 31 feet wide. Their top speed was 35 knots, and they had a crew of 114. The armaments consisted of four 4-inch deck guns and twelve 21-inch torpedo tubes; some had two depth-charge racks on the stern. These were the destroyers that were famous as the ships given to Britain in the Lend Lease deal just before the United States got into World War II. Ships of this class served in other nations such as Canada and the Soviet Union. Others were converted to minesweepers, minelayers, seaplane tenders, and banana boats.

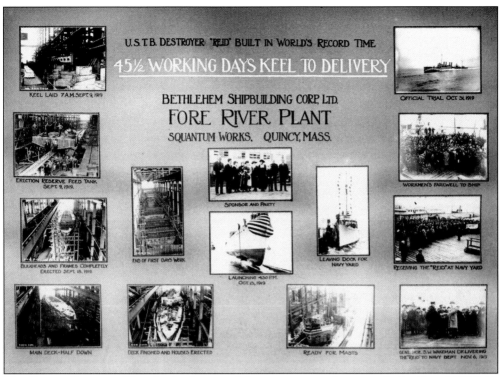

The Victory Destroyer Plant broke the record for constructing destroyers by building USS *Reid*, which was produced in 45 1/2 days. The Victory Destroyer Plant, coupled with the Fore River plant, produced 71 destroyers in 27 months, a record of justified pride.

In this view, the smoke makes USS *Reid* easy to spot. Launched on October 15, 1919, the vessel was named after Samuel C. Reid, master of the privateer *General Armstrong* during the War of 1812. *Reid* was sold for scrap on January 17, 1931.

This view shows USS *Dale* (DD-290) in the left background, USS *Converse* (DD-291) in the center, and USS *Charles Ausburn* (DD-294) right next to the plant, being fitted out. USS *Dale* was launched on November 19, 1919. Richard Dale was captured by the British but escaped and joined John Paul Jones in France, becoming the first lieutenant of the *Bonhomme Richard*. After its peacetime career, *Dale* was sold on January 17, 1931, to be scrapped. Like USS *Osborne*, *Dale* had its hull converted into a banana boat by the Standard Fruit Company of New Orleans. It was renamed the *Masaya* and became an army transport in 1942. The plan was to supply Corregidor, but by the time it and its sister ships arrived, the island had fallen. It spent its army time transporting men and supplies in the Pacific, before being attacked and sunk off the east coast of New Guinea on March 28, 1942, by Japanese dive-bombers. There were 11 fatalities, but most of the personnel on board were saved. *Ausburn* was launched on December 18, 1919, and named after Charles L. Ausburn, who manned the radio on the army ship *Antilles* as it sank. He stayed at his station and went down with his ship. USS *Charles Ausburn* became a pioneer in naval aviation when it was equipped to carry a seaplane in 1923.

Three

THE SQUANTUM NAVAL AIR RESERVE BASE

According to Quincy historian H. Hobart Holly, the idea of using the land at New Squantum as an air base preceded its use as a destroyer plant: "It had served as a training field for a small group of young men who foresaw the important part that aviation would play in the imminent war [World War I]. A small wooden hangar had been erected." That hangar, constructed in 1917, remained in 1923, when Lt. Richard E. Byrd announced that the first naval air reserve unit in the United States would be formed at Squantum. Long before he gained fame as a polar explorer, Byrd served as the first commanding officer of the one-hangar, one-seaplane reserve training facility.

The facilities at Squantum were not designated as a naval reserve air base until 1925, although training had been taking place there for two years by that time. Improvements to the old Victory plant facilities continued into the 1930s until sights like this one, showing crews preparing a couple of N2S Stearmen for takeoff, became commonplace. The Stearmen Kaydets, as they were officially known, worked through World War II and later became famous as crop dusters and barnstorming planes.

The old wet slips for seaplanes used during the Victory plant years provided instant homes for seaplanes like the Naval Aircraft Factory PN-9, an early flying boat. The PN-9 became famous for its ill-fated attempt to fly from San Francisco to Hawaii. When the plane ran out of fuel (as the flight planners had miscalculated fuel consumption and tailwind speed), the plane ditched and, with a jury-rigged sheet, sailed the last 400 miles to the island of Kauai. The stunt proved the hull's seaworthiness, at least, if not the plane's sustainability in the air.

The men stationed at Squantum selected the image of a witch for their insignia, citing the history of the Massachusetts Bay Colony. The insignia adorned the base newspaper and even the bodies of the planes themselves, as shown here on an N3N "Yellow Peril" trainer.

In 1941, when Squantum changed from a naval reserve air base (NRAB), or E base focusing on elimination training, to offering full primary training, 64 N3Ns covered the airfield. The N3N was the last biplane used by the U.S. military and was the last plane manufactured by the Naval Aircraft Factory. Of the 995 built, about 20 remain flightworthy today.

The amphibious Grumman G-21 Goose continues to fly today. Conceived of as an air yacht by a syndicate of New York businessmen, the Goose first saw military action with the Royal Canadian Air Force and later the U.S. Army and U.S. Navy, as well as the air force of Peru and the navy of Portugal. The U.S. Navy and Coast Guard used the Goose in utility, transport, and antisubmarine duty, the latter an important role of the crews at Squantum.

The F6F Grumman Hellcat made a splash when it first arrived on the navy scene in 1943. In fact, it caused several splashes in the Philippine Sea. During the battle there, known as the Great Marianas Turkey Shoot, American forces shot approximately 330 Japanese planes out of the sky and lost only 30 American aircraft. The Hellcat shown here at Squantum is the F6F-5N.

If one could not get close enough to this formation of FG-1D Corsairs, manufactured by Goodyear, to read the word Squantum under the tail section, then all one had to do was read the large letter Z on the tail to know where these planes had come from. This image, a classic shot of classic planes, was used in airshow posters a half-century after the picture was taken. The Corsair was unique among American fighter planes of World War II for its inverted gull wing design, developed to give the plane a lift at the nose to accommodate its oversized propeller. Although a highly successful plane (used first by the Marines in the Pacific and later by the navy once they straightened out a carrier landing problem), the Corsair had its share of frustrations. For instance, the long nose blocked the pilot's view during takeoff. For this reason, the plane earned the nickname "Hose Nose," to go along with "Bent Wing Bird," "Hog," and "Ensign Eliminator," which it gained due to its reputation for those dangerous carrier landings.

Looming over the heads of the Squantum championship baseball team (note the trophy being held at the center), the PBY Catalina was the most extensively built amphibious aircraft in history. Aircraft manufacturers Consolidated and Douglas responded to the navy's 1933 request to produce a long-range patrol plane with heavy cargo-carrying capability, and five years later the navy adopted Consolidated's design as its flying boat of the future. Almost instantly, in 1939, the service began looking for its replacement, but massive orders from allied countries at the start of World War II prompted the navy to give the aircraft a second look. Instead of scrapping it in favor of a newer plane, the navy placed its largest order for any plane since World War I. Used by armed forces around the globe, the Cat played many roles in many battles. This particular plane, a PBY-5A, was given a retractable tricycle undercarriage for use in amphibious operations. Slow and unarmored, the Catalina was vulnerable when attacked, but it carried out nighttime raids for the navy and performed superbly in search-and-rescue operations.

Squantum's years as an active naval air station (officially designated as such on March 5, 1942, and closed on December 4, 1953) coincided with the navy's experimentation with the helicopter. Lt. Comdr. Frank Erickson, U.S. Coast Guard, recommended the navy pursue the development of rotary wing aircraft for the purposes of antisubmarine warfare and lifesaving, after witnessing a test of an early Sikorsky helicopter in June 1942. A month later, the U.S. Bureau of Aeronautics ordered four helicopters for evaluation by U.S. Navy and Coast Guard personnel. On May 7, 1943, navy representatives witnessed a landing exhibition aboard the merchant tanker *Bunker Hill*, and a month later Erickson reported that in his opinion helicopters should be developed as "the eyes and ears of the convoy" on antisubmarine patrol and not as killing machines. On January 1, 1944, the navy began training pilots, and two days later a dramatic rescue mission flown by Erickson himself—delivering blood to a snowbound hospital when all other aircraft were grounded by the storm—convinced all of the helicopter's use as a lifesaving tool. The rise of the helicopter meant the demise of the lighter-than-air ship's service in antisubmarine warfare. Pictured behind these unidentified men are HTE-2 Hiller helicopters.

Few things can keep a Texan down, but put enough snow on the ground and you can keep a North American T-6 Texan from flying, as shown here in a wintry scene at Squantum Naval Air Station. Navy personnel knew the Texan by its designation SNJ or as the Pilot Maker for its important role in training young aviators for combat in World War II.

On March 3, 1948, those same cold temperatures grounded the PBY-5A and the R4D at Squantum. More than 10,000 R4Ds (the navy designation for the DC-3) saw service with the various branches of the U.S. military during World War II, only 570 of which were used by the navy.

The letter Z on the tail signifies that this Grumman AF-2S Guardian belonged to the navy's Squantum station. The navy hurried Grumman's plans to develop a hunter-killer carrier-based antisubmarine aircraft to replace the TBF/TBM Avenger as the war came to a close. The Grumman plant turned out 387 Guardians, most of which were phased out by the mid-1950s. The plane fit the profile of the Squantum's sub-hunting needs.

This early airshow photograph shows the Blue Angels with their F9F-2 Panthers preparing for their routine. The Blue Angels used this particular plane in 1950 and from 1952 to 1954, but its most effective uses were in combat. During the Korean War, Panther pilots were the first carrier-based warriors to engage the enemy, the first to shoot down a Yak-9 and a Mig-15, and the first to complete an air-to-ground attack.

British markings on the wing of the Grumman Goose in the left foreground reflect a definite moment in time for the base. From May 1943 to July 1944, the British Fleet's Air Arm (FAA) trained at Squantum. The FAA had seven Avenger squadrons at Squantum and one Helldiver unit.

The British brought their Fairey Swordfish, a surprisingly effective torpedo-bombing biplane, to Squantum to train during that period. The planes saw action in Norway and Italy early in the war and were equipped with radar for sub hunting as early as 1940. Just a month after the British left Squantum, the last Swordfish rolled off the assembly line, in August 1944.

So what do you do when Mother Nature tells you not to fly? You flee. In this case, many of the men on duty at Squantum fled the South Shore for the relative safety of Niagara Falls, New York, in the face of a hurricane heading up the East Coast. This photograph comes courtesy of Bill Horsch, standing just left of the Centre Street sign with his hat barely clinging to the back of his head. Today, a half-century beyond his duty at Squantum, he is extremely active in the movement to create and open a museum dedicated to keeping alive the history of both the Squantum and South Weymouth naval air stations. He contributed greatly to the production of this book, both with imagery and anecdotes.

When the occasion presented itself, crews from naval air stations joined crews of carriers for flight training. Here, men from Squantum pose aboard USS *Monterey*, a World War II battle-tested carrier serving out its later years under the Naval Training Command at Pensacola, Florida. The visit to the carrier would have been an anticipated break from routine at any air station.

Planes and crews from various naval air stations around the country convened and trained on *Monterey* at any given time. Here, a General Motors TBM-3E Avenger lands while another diverts. The Avenger earned its name for its quick appearance on the navy scene after Pearl Harbor, in January 1942.

The unsung heroes of all aircraft squadrons were the men who kept them in the air. Even while aboard *Monterey*, Bill Horsch and his comrades in the Squantum Avionics Division took advantage of training opportunities to keep their repair skills sharp.

The Grumman F6F Hellcat proved to be one of the navy's most effective carrier-based fighter planes. On *Monterey*, the men of Squantum got to see that fact tested. Some of the men, of course, could have proven that in combat a few years earlier.

Life at Squantum Naval Air Station was not all work all the time. The large concentrations of fit young men on air bases and other military installations naturally led to the development of intra- and inter-base sports leagues. Here, members of the Squantum Avionics Division pose with their baseball trophy in December 1951.

The Squantum Avionics Division chiefs made the statement that short ties were in on the East Coast in 1950. Shown in this image are, from left to right, Frank Haywood, Tom Perry, Jack "Smoky" Miller, and Bill Horsch.

Camaraderie was a natural outgrowth of military life. One could either become friends with the men living in the barracks or be forced into an uncomfortable life as an outcast. Such friendships were hard to part with when war claimed lives and tore comrades apart. Here, at Squantum, Jim O'Connor (seated, left) shares time with some of his friends.

The military has long used civilians in nonmilitary roles on its bases, and Squantum was no exception. This woman, Beatrice (Ramsden) Robertson, worked in the administrative office of the base and told us years later that she chose the lower-paying job over working at a bank.

49

Advancing in rate could be a dangerous thing if one had a lot of friends on base. Here, the men of Squantum demonstrate the proper technique for dunking someone who has done a good deed. First, remove the shoes of the offender and grab him by both the arms and the legs.

While tossing the man into the water, throw your arms in the air and be sure to holler, "And stay out!" Add as much body English as you feel necessary.

Be very careful, though, during the dunking that you are not standing too close to the water's edge. Your friends will find you as a tempting target.

When coming ashore, be sure to get everybody wet, lest they all miss out on the real fun of a dunking.

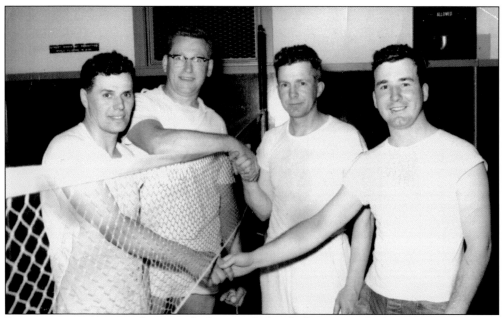

Recreation came in many forms on the base, including badminton and other competitive sports such as bowling and football. In this 1948 photograph, M. Jesus (far left) and John Novick (second from left) shake hands with Joe Sloan (third from left) and Tom Lynch.

Recreation could mean occasionally tapping a keg at Houghton's Pond, as shown here in the summer of 1949.

The most dangerous achievement for any man in the navy may have been the advancement to chief. Here, Lou Gribling deals with his chief's ceremony, as administered by Bill Horsch (left) and Frank Bourdon (right). By tradition, the new chief was given a trough and was spoon-fed various foods, some of which would seem to be quite unpalatable to the normal consumer. In fact, if the new chief felt that way, he was given a choice. He could eat the food or wear it. Horsch remembers his ceremony this way: "When they came out with a bucket of raw eggs, and said 'Eat it or wear it,' I said, 'No way! I'll wear it.'" His servers poured the bucket of eggs over his head. Nobody ever said navy life would be easy.

In 1953, with increasing air traffic in Boston Harbor, as the navy competed for air space with the growing Logan International Airport, the government opted to move its operations inland to South Weymouth, reviving the old lighter-than-air station opened in the early 1940s. Developers pushed forward ideas for new uses of the land, including civilian airstrips and this dramatic representation of a Jordan Marsh shipping and storage facility.

Today, barely anything remains of the old Squantum airfield, now covered with the condominiums and restaurants and the shops of Marina Bay. The only outward sign of any military significance at all is the Vietnam Veterans Memorial, shown here. Thousands of men and women worked here for more than three decades, helping in two world wars, but in the end, their legacy is forgotten by most who visit the site.

There were two blimp hangars built for the blimp squadrons at Naval Air Station South Weymouth. Shown here is the construction of Hangar 2, which was built of wood instead of steel. The doors were hung in large concrete structures. Their frames are seen being built in this photograph. Joseph Walen bought a truck like the trucks seen here and worked on the construction of the base. He could not identify his truck in this view. He later traded the truck for approximately 50 acres of land in Rockland and went back in the pig business.

This is the staging for the concrete doorframe for Hangar 2. This hangar was started in 1942, and because steel was needed elsewhere, it was constructed of wood.

This photograph, dated April 4, 1942, shows the doorframes are complete on the far end of the hangar and the arch roof and sides are progressing on Hangar 2. All of the lighter-than-air hangars constructed after the war were built of timber. The book *Building the Navy's Bases in World War II* describes the construction of the hangars this way: "The timber hangars consisted of a shell, which was half-egg-shaped in cross-section, and two end doors. The shell was 1,000 feet long, 196 feet wide, and 170 feet high, supported by a series of 51 transverse timber arch bents, and wood sheathed. The lower 24 feet of the shell was framed by a series of reinforced-concrete bents which transferred the arch loads to the foundations."

By the 1960s, the original Hangar 1 had outlived its usefulness. The blimps were gone, and in 1967, the old hangar was taken down. Nevertheless, a hangar for fixed-wing planes was needed. This is the new hangar being constructed in 1967 at the same site as the old Hangar 1.

The concrete-arch beams being put in place in this photograph weigh 60 tons. On August 18, 1967, seven of these arches collapsed, killing two workers, Sabatino Rasetta and William McGowan. There were also five workers who were injured. This hangar was finished in 1970.

This photograph gives a good perspective of the internal structural support of the wall and the roof of Hangar 1. There was a walkway built into this in order to allow workmen to climb to the top of the hangar. One gets an idea of the height from the man standing on the ground in the picture.

At Naval Air Station South Weymouth, this was the top of the world, the view from the roof of Hangar 1. From this vantage point, one could easily see tall buildings in Boston to the north and the sea to the east. This structure could also be seen from points in nearly every South Shore town.

This side view of Hangar 1 (still under construction) shows the two-story lean-tos set into the side of the hangar. These structures were used for officers, classrooms, and shops. In this wartime photograph, the base water tower and the hortonsphere (the sphere that held the helium for the blimps) have been constructed.

This is a side view of Hangar 1 with its lean-tos on the side. This is a good view of the hortonsphere as well.

Hangar 1 was a marvel and was constructed to hold six blimps. It was made of steel and covered eight acres. The dimensions were 956 feet long, 294 feet wide, and 191 feet high. The doors were known as orange peels and weighed 250 tons each. It took four large electric motors to open and close each door along railroad tracks. This made it one of the largest hangars of its type in the world. The K-type blimp that used the hangar was 425,000 cubic feet large and 250 feet long. Construction of this hangar began in July 1942, and it was finished in November 1942. Beside the two blimp hangars, the base provided a circular mat of 2,000 feet, a grass runway of 4,500 feet, as well as six mooring circles 500 feet around.

Five
SOUTH WEYMOUTH DURING WORLD WAR II

This aerial photograph shows Naval Air Station South Weymouth as it looked during World War II. The long building to the upper left is Hangar 2 with a blimp moored in front. The building in the center is Hangar 1 with the east mat below it. The circular objects to the left hold radio antennas. The one with spokes could also serve to calibrate a compass. To the right of Hangar 1 is the rest of the base with the offices and barracks.

This is Hangar 1 with its doors open and two blimps inside. The residents of the Massachusetts coast had experienced German U-boat scares during World War I. On July 21, 1918, *U-156* shelled a tug and its four barges off Cape Cod. With the anticipation of America being involved in World War II, a maritime defense against the U-boat was of utmost importance.

This is a ground view of Hangar 1 with two blimps on portable triangular mooring masts. You can gauge the height of the mooring mast by the lean-tos on the side of the hangar, which are two stories high. The standard mast was 42 feet high. Goodyear made these K-class blimps.

This is a beautiful photograph of a K-class blimp and part of its standard 40-man docking crew. It is hard to say if they are walking it in or out of the hangar. The tail fins of the K-class blimp are set at right angles. Most of the post–World War II blimps had their fins at a 45-degree angle. During the war, flight crews usually consisted of 10 men. The officers were a pilot, two copilots, and a navigator. The crew was made up of an airship rigger, an ordinance man, two mechanics, and two radiomen. The blimps were armed with four 350-pound depth bombs and a .50-caliber machine gun. Other arms included automatic rifles that could be installed on mounts at various windows. The blimps had radar that could detect a ship 90 miles away as well as sonobuoys and other underwater detection equipment and navigation instruments for day and night flying. The blimps from South Weymouth used this equipment in the search for German U-boats along the Northern Atlantic Frontier.

These K-class blimps were the workhorses of the first naval district effort in maritime patrols, especially when escorting convoys. The section of the eastern sea frontier that was under the first naval district extended from Mount Desert Rock to Nantucket Shoals. The runs between Boston and Halifax were the main convoy routes for the blimps of South Weymouth.

On September 2, 1943, the commander of the Eastern Sea Frontier (which extended from Canada to Jacksonville, Florida), Vice Adm. Adolphus Andrews, made an official visit to Naval Air Station South Weymouth. He is the one standing in front of the ladder. The only other person identified in this photograph is the man in the dark uniform, Capt. V. E. Hebster (U.S. Navy, retired), subcommander of the northern group. Unfortunately, the officers chose to pose in front of the open head, or restroom.

When severe storms threatened the coast, planes and blimps alike sought shelter and safety. This photograph shows about 90 fixed-wing planes and blimps all sheltered in Hangar 2 during a storm. Most likely, it is the storm of September 14, 1944. At this time, the planes had to land on a grass runway. During the war, other innovations were part of base life; pigeons were used on blimps in order to send messages without breaking radio silence. One story relates that the pigeons, instead of flying back, landed on top of the blimp and rode back. In another innovation, women were hired as security police and armed with shotguns, due to a shortage of able-bodied men.

This is K-38 making an emergency landing. It looks like there may be more than the usual 40-man landing crew on the ground. Note the lines on the blimp. High winds could be dangerous for a moored blimp. In the summer of 1944, the K-6 blimp was moored when a small tornado tore it from its mast and blew it into the woods. In the air, the K-class blimp was known for its endurance. It could cruise for 26 hours at slow speeds at altitudes of 100 feet. Also, it could be operational when other aircraft were grounded because of poor visibility. These blimps proved themselves in antisubmarine warfare. Blimp Patrol Squadron 11 (ZP-11) at South Weymouth was proud of the record of its eight airships.

This is K-11, which was patrolling with two other blimps over the South Shore when they were caught in a thunderstorm on July 21, 1943. The others headed out to sea, but K-11 was caught in a downdraft and crashed in Scituate near Maple Street. It hit a barn, which was flattened, and the bag was caught in a tree. The crew members were unhurt. This is one of only three major blimp crashes during World War II. In fact, there were three total crashes for the first naval district.

This 1940s photograph shows the administration building of the base, and behind it is the original fire station. The last building before the water tower is the power plant.

This is the bachelor officers' quarters. The photograph may have been taken from a blimp. The car may be an official vehicle on base business. The cause of the tracks around the trees is unknown.

This chapel did not appear at South Weymouth until 1962. It originated in the 1930s at Camp Edwards on Cape Cod as two army regimental chapels that had been abandoned. It took from the summer of 1959 to September of 1962 to move, assemble, do the paperwork, and dedicate the new Friendship Chapel at South Weymouth.

This is the old post exchange and special service building. The photograph was not taken before 1976, for that was the year the triangular station logo replaced the old shield-shaped logo. This building was destroyed by fire on February 18, 1979.

Across from the bachelors officers' quarters was the housing for chief petty officers, a series of duplex houses. In the 1950s, William A. Horsch of Rockland and his family occupied the house that is second from the right in the photograph.

This is *U-1228*, a IXC-type U-boat. It was a long-range submarine with a snorkel, and its operational area was to be off the coast of New York when the war ended. This photograph was taken on May 17, 1945, after *U-1228* had surrendered and was being escorted into the navy yard at Portsmouth, New Hampshire. The U-boat was defeated in World War II, and much of the credit goes to the blimps of South Weymouth.

Six
FROM V-J DAY TO DECOMMISSIONING

At the end of World War II, Moscow was the place where decisions were being made that affected the operations and mission at Naval Air Station South Weymouth. These two blimps are ZPG-2 N-class airships developed to act as early-warning craft against Soviet planes and submarines. The Naval Air Development Unit (NADU) was established in July 1953. A blimp named *Snow Bird* made a record-breaking trip of 9,448 miles, captained by Ronald W. Hoe, commanding officer of the NADU. The last blimp flew out of South Weymouth on August 2, 1961.

Shown here is a change-of-command ceremony in the 1950s, inside Hangar 1. The mammoth hangar made a large gathering look small. Condensation in the hanger fell like rain, hence the expression "the hangar had its own weather." It was in 1953 that the naval air station at Squantum was closed and its command was transferred to South Weymouth.

By the time this picture was taken in the 1950s, fixed-wing aircraft were taking up space in Hangar 1, along with the change-of-command ceremony. The plane in the foreground is an R4D Skytrain. The letter Z on the tail represents Naval Air Station South Weymouth, the designation transferred from Squantum.

This is Hangar 1, and the date of the photograph is August 31, 1954, when Hurricane Carol, a category 3 hurricane, came ashore in southern New England. This hurricane sustained winds of 80–100 miles per hour and was the most destructive storm since the Hurricane of 1938. All the aircraft at Naval Air Station South Weymouth were gathered up and put into the hangar for safekeeping, including the two N-class blimps, the type used by the Naval Air Development Unit. By the number of fixed-wing aircraft in this photograph, it is easy to see that the lighter-than-air operations have been superseded by the airplane.

Shown c. 1952 is the first helicopter at South Weymouth. It is a Helo, HTE-2 made by Hiller and used for training and utility purposes. Antisubmarine helicopters were attached to the base until the late 1960s. The SH-3A Sea King helicopter was one of the last to be used for this purpose at this base.

This is a 1946 photograph of 1180, 1188, and 1200 North Union Street in Rockland. The houses in this photograph were either moved or taken down for the 1958 naval air station runway expansion. This section, which borders Weymouth and Rockland, is rich in early history. The first settlers moved south from the Boxberry section of Weymouth to what was East Abington in the 1730s.

The first house on the left was 1180 Union Street in Rockland and is believed to be the shoe factory of Hunt and Loud, who made commercial shoes c. 1825. It was Capt. Thomas Hunt who started the shoe business in this area of Union Street in 1793. That business led to the area's biggest industry. This is an early view of the same area as the previous photograph.

This is 1281 Union Street in Rockland, believed to have been Civil War veteran David Whitman's house. About 20 homes were removed or razed for the expansion of the runway across Union Street in 1958. The north end of Union Street in Rockland to Union Street in Weymouth was closed to build the base. The *Brockton Enterprise* suggested that a tunnel be dug under Union Street so traffic would not be have to be rerouted.

This 1941 photograph of 1300 Union Street in Rockland, bordering Weymouth, was the home of David Whitman Jr. (1780–1832). This house was moved to Circuit Street in Hanover c. 1960. This area of the base is rich with history, including early American Indian trails. The Massachusetts Bay Colony and the Old Colony of Plymouth set their boundary here with the Accord of 1664. Early residents of Weymouth and East Abington (Rockland) settled here in the Old City and Boxberry areas, and the local shoe industry started here.

This easterly view of the construction of the runway extension was photographed in 1958. Union Street runs across the image from right to left. The 1950s were a busy time for the defense department as the cold war was reaching its peak. The United States was engaged in armed conflict in Korea between 1950 and 1953. In 1950, plans to reopen Naval Air Station South Weymouth were under way. In 1953, Naval Reserve Air Base Squantum (NRAB) was closed and Naval Air Station South Weymouth reopened. Given the transfer of both U.S. Navy and Marine Reserve units and squadrons, and the requirements of modern aircraft such as jets, many improvements had to be made to the base in order to accommodate their missions.

Looking east, this 1958 view shows the expansion of the east–west runway. Most of the trees have been taken down, and the edge of the construction is clearly defined. Union Street is shown still intact, running from left to right (north–south) in the image, but only construction vehicles are using it. This extension made the east–west runway 4,000 feet long, with a total length of 6,000 feet. Not seen in the photograph is the north–south runway, which is 7,000 feet long.

This view of the extension of the east–west runway looks south from Weymouth to Rockland. Much of the heavy work has been done, but graders and heavy trucks are still at work. When Union Street closed, it transformed from a busy main street to a quiet neighborhood in this part Rockland as well as Weymouth. People who lived there remember the sudden silence. You can see part of the east blimp pad at the lower right.

This photograph shows the naval air station in the upper left corner, with Weymouth Street and Abington Street. A bypass was constructed to replace the loss of Union Street. In 1959, the federal government spent $922,000 to build Veterans of Foreign Wars (VFW) Drive. This road would also serve to access Route 3, which was being built at the same time.

This photograph was taken in the 1960s. Hangar 1 is still standing and the runways have been extended. The runway extension was finished in 1960, and the hangar was razed in 1966. This aerial photograph gives a pilot's point of view, flying in from the west. After World War II, there was a serious need for housing. This view shows the spread of the housing development around the base in the 1960s. This development continued until the base closed. At the time of its closing, the base was one of the largest undeveloped parcels of land on the South Shore.

These drainage pipes were put in at the time of the extension of the east–west runway. This base was built largely on wetlands; therefore, there was a need for a good drainage system.

These men are installing arresting gear on the east–west runway. Navy planes had tail hooks to catch the cable of the arresting gear, which stopped them when making carrier landings at sea. Tail hooks were standard on land-based navy planes as well. As engines and electronics in planes became more sophisticated, arresting gear remained decidedly low tech.

The arresting gear consisted of strong cable across the runway with large anchor chains along the side of the runway, as seen in this photograph.

This is a closeup of the links of anchor chain used in the arresting gear at Naval Air Station South Weymouth.

Fuel was not much of a concern when the blimps flew at South Weymouth. Later, when jets and other planes flew, aviation fuel was needed and tank farms were established. The fuel was stored in underground tank farms. This is the pumping station.

The structures on the mounds over the tank farm were known as "dog houses." The belief is they were built to disguise the fuel storage from Soviet spy satellites.

This is another view of the tank farm. There are jets parked behind the building on the left. They are A-4 Skyhawks, which were used both by U.S. Navy and Marine Corps squadrons at Naval Air Station South Weymouth.

The new aircraft in the 1950s required new firefighting skills and equipment. This construction is for the new fire station. It is also a good view of the two hortonspheres, which held the helium for the blimps. After Hangar 1 was taken down, the water tower became one of the landmarks of the base.

This aerial photograph dates from the 1980s. The building in the background with the sloping side is the Shea Recreation Center, which opened in 1981. Other new buildings built in the 1980s were the bowling alley and a new main gate. The main gate was originally off White Street, a residential street in Weymouth. After a fatal accident in 1981, it was decided to move the gate to Route 18, where many thought it should have been in the first place.

This c. 1944 photograph taken from Hangar 1 gives a good view of some of the buildings of the base during World War II. The administration building is in the background. The building this side of it is the fire station. The three silos, completed in 1943, are for coal storage. To the right of the silos is the power station that was built in 1942.

Looking north from over Rockland, this is a view of the base at its full development. The Old City was taken when the navy developed the base. The runways will be taken by future commercial development of the base land. Since this photograph was taken, more houses have been built in the communities surrounding the base.

The first plane on the left is Skymaster (R5D). These were transports that replaced the World War II vintage Skytrains (C-47J), which were transferred from Squantum. This aircraft was in service until the 1970s. The plane to the right is a Neptune (P-2V). This is the plane that replaced the blimps on maritime patrols searching for Soviet submarines after World War II. The Orion (P-3A) replaced this plane in the 1970s.

This is the Tracker (S2F) from Naval Air Station South Weymouth. The station designation is clearly marked on the tail of the plane, 7Z. The Z on the tail goes back to Squantum; the 7 was added later. The code represents an aircraft from the reserve training center at South Weymouth. The submarine is American, and this is a training exercise. The work of detecting submarines off the New England coast continued up to the closing of the base in 1997. The work was important and dangerous, but it was done well by the reservists who flew the planes. Squadrons doing offshore patrols used a variety of aircraft. The Neptune (P2V-7) and its replacement, the Orion (P-3), performed this work. There were also helicopter squadrons doing antisubmarine work that were stationed at South Weymouth.

This is the Neptune (P2V), which was used as a land-based patrol plane from the early 1950s to the early 1970s. It was used for antisubmarine long-range patrol and was able to detect and attack submarines.

There are two types of jets in this photograph. The planes toward the back are the Cougars (F9F-6). They were the swept-wing version of the Panther (F9F), a U.S. Navy and Marine carrier plane during the Korean War. The Cougars were produced too late to see combat in the war. The planes in the foreground are Banshees (F2H-2N). They had combat action in the Korean War with the U.S. Navy and Marines. They were made famous as the plane in the James Michener novel *The Bridges of Toko-Ri*.

An Orion (P-3C) is snow covered from the blizzard of February 1978. The storm delivered winds of 78 to 92 miles per hour, and 29 inches of snow in eastern Massachusetts. The personnel and their equipment on Naval Air Station South Weymouth assisted the local towns. The Orions replaced the Neptunes (P2V) as of the mid-1970s. The plane was fashioned on the airframe of Lockheed's airliner the Electra (L188). The mission of the squadrons in using this plane was to search for, detect, and destroy enemy submarines. The plane can maintain a 13-hour mission and is armed with torpedoes, mines, and depth bombs. This plane is in the military inventory of nations around the world and is used in missions over land as well as over the seas. The last squadron of Orions at South Weymouth was transferred to Naval Air Station Brunswick (Maine) in 1996.

The planes parked in front of the new Hangar 2 are Skyhawks (A-4).

Two attack squadrons were at Naval Air Station South Weymouth in the 1960s and 1970s. Skyhawks (A-4) were used to train naval pilots. This type of training unit had a tradition that went back to Squantum. This all ended in 1992, when the last attack squadron was deactivated at South Weymouth.

Marine attack squadrons were attached to Naval Air Station South Weymouth after they were transferred from Squantum. This type of squadron's mission was to give close air support to ground troops. The last plane used in these units was the Skyhawk (A-4 M). The Skyhawk was developed in the mid-1950s as a lightweight carrier bomber. This plane made the first carrier raid on North Vietnam in 1964. The Blue Angels used this plane in their demonstrations from 1974 to 1986. In the late 1970s, it was replaced in carrier inventories and began to be employed in reserve units. The last of the Marine attack squadrons was deactivated in 1992. The Skyhawk was the last jet station at Naval Air Station South Weymouth. The plane that was made into a monument and located at the main gate is a Skyhawk. Capt. Ray Demming of the U.S. Navy acquired it from the naval training center in Memphis, Tennessee, in 1980. It was the same plane he flew when he served in the Naval Air Station South Weymouth Navy Attack Squadron.

It was the skill of the welder that built the hortonspheres in the 1940s, and it was the skill of this man and his torch that disassembled the spheres in the 1960s.

Sadly, what was once a necessity and a great marvel became too large and costly to maintain. In this 1966 photograph, Hangar 1 is being dismantled. A smaller hangar was later built on the site.

The sheeting has been removed from the great doors of Hangar 1. You can see the lean-tos on the far side of the hangar.

Near the end, there are only a few arches of Hangar 1 remaining.

The lean-tos stand alone, except for one small section of the wall support on the far end of Hangar 1. The new hangar was built between these lean-tos. The old water tower was taller than the new hangar. The new hangar was still standing in 2004.

The last arch is being taken down, and the official delegation is there to witness it. They are standing in a group in the foreground of the photograph. We do not know for certain who is in the group, but we believe the base commander is there. In 25 years, Hangar 1 at Naval Air Station South Weymouth went from great achievement and a South Shore landmark to a memory. The base closed 30 years later. This ended 200 years of naval base history in the Boston area, starting with the Boston Naval Shipyard, established in 1800.

Decommissioning Crew

Know All Ye by These Presents and to all land lubbers, sea lawyers, salts, swabs, squareknot admirals, goldbrickers and other scavengers of the seven seas *Greetings:*

Know Ye That Lawence M. Piller

1942

1997

was an honored member of the illustrious crew which forever distinguished itself when it decommissioned Naval Air Station, South Weymouth, Massachusetts, and therefore, for this good and sufficient reason, is entitled by the laws of the sea to all rights and privileges afford by the demise of such an exhaulted organization and is hereby granted clear, free, open and uncumbered dominion over the morass of scoundrels departing their post prior to last muster. Dominion is to be exercised according to the honored rights of seniority according to the treasured records contained in Davy Jones' Log. Disobey this order under penalty of our royal displeasure.

Decommissioned
30 September 1997

R.A. Duetsch, Captain, USNR
Commanding Officer

This is the certificate that the last crew members received when Naval Air Station South Weymouth was decommissioned in 1997.

Seven
South Weymouth Airshows

Most residents of the South Shore will remember two things about Naval Air Station South Weymouth: the omnipresent hangar looming over the treetops, and the spectacular airshows put on to entertain the public over the years. As thousands gathered to look skyward—as had their ancestors at the Harvard-Boston Aero Meets of 1910, 1911, and 1912—the Blue Angels and other stuntmen and precision flying aircraft took to the air, allowing the workhorse P-3 Orions and the other standard aircraft of South Weymouth the opportunity to stand down for a few days.

Coordinating the airshows was a monumental task; the base's recreation staff had to plan on not only feeding and hydrating thousands of spectators but also finding enough room for them to park without letting them get in the way of the aircraft participating in the show. Airshows are nearly as old as manned fixed-wing aviation itself. The first international airshow took place at Belmont Park Racetrack on Long Island, New York, from October 22 to 30, 1908. At that show, 27 airmen from three countries showed their stuff. Five years later, on September 22, 1913, flyer Adolphe Pegoud turned his plane upside-down, becoming the first man ever to do so, and a few weeks later turned the world's first loop-the-loop. The coming of World War I forever changed the general perception the public held about aircraft. Aircraft were once novel inventions of limitless and wondrous possibilities, but the war turned them into killing machines. Pilots discharged from military flying at the end of World War I formed the class of pilots known as barnstormers, traveling the country putting on exhibitions of the proficiency they had gained during the combat years, and for a short time bringing back the popularity of exhibition flying.

The traffic flow into or out of the South Weymouth base could be a nightmare since only the new access gate off Route 18 was open for spectators.

The recreation staff at the base tried to find something to capture the attention of everybody who attended the airshows. Here, Wally the Whale, chosen as a mascot because of Massachusetts' legendary links to the sea, entertains some of the younger members of an airshow crowd.

Once all of the spectators had arrived and found their way to the airshow field, the planes began their ascents, like this deHavilland Chipmunk. Known as the Pepsi Skydancer, it was built for the Royal Canadian Air Force in 1956 as a trainer and later converted for aerobatics.

While there really could not be a bad spot from which to view the show from anywhere on the field, there is no doubt that access to the roof of the hangar would have been a coveted privilege for spectators.

From the roof of the hangar, the view of the Grumman cats and all of the other planes must have been tremendous. Grumman's famed fighter planes starred in World War II. The F4F Wildcat held off the Japanese during the first year of the war as a frontline fighter and was joined in 1943 by the F6F Hellcat. The F8F Bearcat, delivered just a few weeks before the end of the war, never made it into combat.

Once the pilots got warmed up, the show really began. Here, the navy's Blue Angels, in their FA-18 Hornets, swarm in the sky. The speed and power of modern aircraft call for the ultimate sure-handedness and concentration on the part of the pilot under normal conditions; precision flying and aerobatics bring that concentration to another level.

The airshows offered a chance for spectators to see living history as the planes of yesteryear shared the skies with modern craft. The Consolidated PBY Catalina was the answer to America's need for a long-range patrol-boat aircraft in the late 1930s and served with distinction throughout World War II in almost every theater of action.

The Douglas DC-3 played many military roles during World War II, but its role as a civilian people carrier may be its greatest legacy. As a military transport, the DC-3, or C-47, carried several official names, such as Skytrain, Skytrooper, and Dakota, and one unoffical, "Gooney Bird."

Had they met in another time, this Mig 15 and P-51 Mustang would probably have squared off in combat. Instead, they shared airspace at airshows in the 1990s. The Mustang, *Crazy Horse*, was built in 1944 as a P-51D and was later converted to a TF-51 dual cockpit, dual-controlled Mustang, becoming one of only twelve in the world.

Unfortunately for the Mig, its visit to Naval Air Station South Weymouth resulted in a botched and disabling landing, making it look as if it had been in combat. Mig 15s first appeared after World War II. They were manufactured by the Soviet Union and sold or leased to satellite communist countries. American pilots fought against Mig 15s in the Korean War.

No airshow in America is complete, it seems, without the internationally known Gordon Bowman Jones behind the microphone. Respected worldwide for his knowledge of aircraft of all types and their particular safety concerns, he not only announces shows live and on television but has also been appointed by the Federal Aviation Administration as a safety inspector for its aviation safety program. Jones called the shows at South Weymouth in the early 1990s.

Jones saw many oddities in his time, including this U.S. Navy balloon juxtaposed with the U.S. Army DC-3 *Precious Cargo*. The military of the United States began using balloons as early as the Civil War for combat surveillance, harnessing the lighter-than-air concept well before the first fixed-wing aircraft took to the sky at the beginning of the 20th century.

Early pilots did not wear parachutes, thinking them unmanly and symbolic of cowardice. Keener minds eventually prevailed over that logic, though, and the parachute became standard issue of American airmen. In time, jumping out of a plane with a parachute went from cowardly to sporting, as today thousands of Americans participate in skydiving. This jumper made a patriotic statement during a time of political unrest in the early 1990s, as America headed back to full-scale war for the first time since leaving Vietnam in the 1970s.

Airships like the Gulf blimp brought back the nostalgia of the short two-decade span when lighter-than-air craft ruled the skies over South Weymouth station. The blatant commercialism of this airship stripped away most of the warm feelings of the past for the spectators, leaving them with the subliminal notion that as soon as they left the airshow they should buy gas.

Approximately 40 countries around the world use the Bell UH-1 Iroquois helicopter, better known as the Huey. Since the 1950s, more than 9,000 Hueys have been manufactured, and the first American versions saw combat in Vietnam in 1963. Flexible to fit all sorts of mission requirements, from combat to medevac cases and more, it may be the most widely used helicopter in the world.

The spectators at South Weymouth were treated to more than just a flying exhibition by the Hueys of Marine Light Helicopter Squadron 771. The unit, based at South Weymouth from the 1950s until its deactivation on August 1, 1994, showed off its rapid insertion techniques with Marines rappelling their way to the ground and then reboarding the aircraft.

Inglorious in aesthetic design yet functionally superb, the CH-54 Skycrane served as a workhorse and pack animal during the Vietnam War. Designed by Sikorsky, the Skycrane featured a hoist that allowed its crew of three to pick up or deliver cargo without landing. That cargo could have been a disabled aircraft or anything too big for its sky-going partner, the CH-47 Chinook, to handle, including a 10,000-pound crater-making bomb used to create landing zones in dense jungle, a 45-man people pod for delivering troops to combat, or even a mobile army surgical hospital (MASH). The hospital came with an x-ray, blood lab, air-conditioning, and proper lighting for emergency surgery. The Skycrane served with the U.S. Army's 1st Cavalry Division in Vietnam between 1964 and 1972 and eventually saw service with the Army National Guard. The last active Skycrane retired in 1993 in Reno, Nevada. To see a Skycrane would have been a special occasion for any airshow spectator, as the army received delivery of only 97 of them during their combat in Vietnam, as compared to 5,000 Hueys.

Not all of the aircraft at the shows were strictly military. Pilot Sean D. Tucker regularly wowed crowds with his stunt flying and still does today. Once afraid of flying, Tucker took his apprehensions head-on and became one of the most respected stunt flyers in the world. Here, he performs a ribbon-cutting stunt, flying inverted down the runway. Tucker today is the Federal Aviation Administration's designated ACE (airshow certification evaluator).

Can a P-3 Orion outrace a car built to break speed records in the desert? The recreation staff at South Weymouth sought the answer by arranging for a race between the two, dubbing the event "Smoke and Thunder." With its parachute deployed and well in the lead, it looks like the car won this round.

Even the U.S. Air Force got its foot in the door of airshows at Naval Air Station South Weymouth. The Thunderbirds, the air force's official demonstration team and 3600th Air Demonstration Unit, came together on May 25, 1953, just six years after the founding of that service, which separated from the U.S. Army in 1947. Based at Luke Air Force Base in Arizona, the Thunderbirds derive their name from southwestern American Indian mythology and a Thunderbird—some say an eagle, some say a hawk—that rose from the desert shooting lightning bolts from its eyes. The Thunderbirds have used the F-84G Thunderjet, F-84F Thunderstreak, F-100 Super Sabre, F-105B Thunderchief, F-4E Phantom II, T-38A Talon (a trainer, during the fuel crisis of the early 1970s), and the F-16 Fighting Falcon as their signature aircraft. The latter craft, pictured here, is the longest lasting of all of the aircraft, having been the choice of the team since 1983, in both the F-16A and C models. At least once, they roared over South Weymouth.

An airshow in the early 1990s brought the planes that Americans had seen on television and featured in newspapers during the Gulf War up close and personal. And, if the movie *Top Gun* did anything for America, it introduced the average citizen to the F-14 Tomcat, as well as actor Tom Cruise. A capable air-to-ground, air-superiority, and fleet-protection strike fighter, the Tomcat is slowly being phased out of service in America for the F/A-18 E/F Super Hornet.

The long-range vision of Juan de la Cierva, the inventor of the autogyro, eventually came to fruition with the development of the AV-8B Harrier II, a vertical takeoff and landing airplane. The Harrier II saw heavy action in Operation Desert Storm as the first Marine Corps tactical strike aircraft into the theater of operations. Based only 35 miles from the Kuwait-Saudi Arabia border, it flew the shortest distance into combat of any aircraft. Spectators gawked as they watched an airplane ascend directly into the sky and hover, defying all preconceived notions of the flight of fixed-wing aircraft.

The Gulf War also introduced Americans to the A-10 Thunderbolt II, or Warthog. This low-flying tank killer was the first plane specifically developed for the U.S. Air Force for close air support of ground forces. Its secondary missions include support of special forces personnel and search-and-rescue missions for downed airmen or other military personnel in need of such assistance.

The C-5 Galaxy is almost a football field long. Stretching 247.1 feet from nose to tail, it is one of the world's largest aircraft. It can carry unusually large loads at jet speeds, transporting troops and equipment around the world on short notice. And, with its nose and rear cargo openings, it can load and unload cargo at the same time. The enormity of the plane marks the growth of the airplane industry—in many ways—since Claude Graham White and his compatriots soared on the flimsy wings of their planes at Squantum less than a century ago.

The Blue Angels have come a long way since they were formed in 1946, inspired through the vision of Adm. Chester W. Nimitz, then the chief of naval operations. From aircraft types to formations, they have adapted over the years to ensure safety in the air, impressiveness for every crowd, and that their message of the importance of naval aviation remains in the minds of the American people. When their transport arrives, spectators know that a spectacular airshow is not far behind.

The first Blue Angels flight demonstration team flew the Grumman F6F Hellcat but quickly changed within the first year to the F8F Bearcat for its power and maneuverability. In 1949, they transitioned into the jet age and have since flown the F9F-8 Cougar, F11F-1 Tiger, F-4E Phantom, and others before settling into today's F/A-18 Hornets, which they have been using since 1986.

Early shows simply consisted of 17-minute routines of V and echelon formations but were soon expanded to include diamond formations and more. As speeds increased, maneuvers became progressively more complicated. Today, six pilots fly 30 stunts in 45 minutes, never repeating a maneuver.

The Blue Angels were scheduled to fly one last time at Naval Air Station South Weymouth on June 8 and 9, 1996, but did not appear after their commander pulled them out of the event, citing his own personal apprehensions about the upcoming flights. Listed as the "Blue Farewell," the airshow, which went on without them, was billed as the goodbye to the air base that had protected the Boston area for more than a half-century.

Unfortunately, the airshow coincided with the graduation ceremonies of Rockland's 1996 high school class. As the sound was sure to drown out speeches and addresses at the graduation ceremony, the crew at the air station sent a peace offering to the local students in the form of a small plane flying a banner behind it, congratulating the seniors on their achievements.

On September 30, 1996, the last operational plane flew away from Naval Air Station South Weymouth, ending a longstanding multigenerational tradition on the South Shore. Machinist Larry Piller, a civilian working at the base, was one of the last people off the field. "It was an awful, awful feeling," he said in an interview. No matter what becomes of the land at South Weymouth, the memories of the base will remain in the minds of the local citizens for a long time to come.

Since the last plane roared off the runway at South Weymouth, the skies over the South Shore have been relatively quiet, broken only by the sounds of the regular schedule of flights out of Boston's Logan International Airport. The closing of the base left many civilian employees looking for new jobs in the area and a large tract of mostly unused land waiting for redevelopment or designation as open space, or both. As developers, activists, open-space committees, planning boards, and conservation commissions continue to discuss the future of the former South Weymouth Naval Air Station, memories of what once was begin to fade. South Shore residents trained to keep their eyes on the sky have found little to look up to in recent years.

For nearly a century, the people of the South Shore towns, especially Quincy and Weymouth, but also Hingham, Hull, Rockland, Abington, Hanover, Norwell and others, had witnessed the growth of America's fascination with powered flight. From the biplanes and triplanes at the field at New Squantum to powerful jets like the Grumman F-14 Tomcat, they had seen the most up-to-date aircraft, both civilian and military, that the country had to offer. With the closing of the South Weymouth Naval Air Station, a chapter of the region's history closed forever.

ACKNOWLEDGMENTS

This book is the work of an extended family of people, rather than just the two names that appear on the title page. Each person or institution listed below significantly contributed to the production of this book in a special, specific way.

Our research began at the Weymouth and Quincy Historical Societies. We would like to thank the volunteers of both organizations for taking their time in answering questions and digging out obscure references we felt necessary for the completion of this work. We particularly extend our thanks to Ed Fitzgerald, director of the Quincy Historical Society, for his time and effort.

Invaluable information was also located at the Thomas Crane Public Library in Quincy and the Tufts Memorial Library in Weymouth. We wholeheartedly thank the reference staffs at both facilities for their help and endorse both institutions as excellent places of learning for our region. The photographic collection stored at the National Archives and Records Administration Northeast Division headquarters in Waltham led us down several unexpected paths, thanks to archivist Joanie Gearin and the NARA volunteers, who always seemed to find us one more thing of interest. Dave Barney and the skeleton crew still working for the U.S. Navy at what was once Naval Air Station South Weymouth provided numerous key photographs and leads to other collections.

Bill Horsch, who appears in this book in his military gear of a half-century ago, provided a wealth of information about life at the Squantum station. Through him we also received access to the collection of Joe Sloan, with whom he served. Barbara Aikens helped us identify photographs of old houses on Union Street, taken from the collections of the Historical Society of Old Abington and the Dyer Memorial Library.

Joseph Walen and Irene (Norkus) Matiyosus explained life in Weymouth's Old City for us, as only local residents could. Larry Piller, also of Weymouth, granted us access to his photograph and video collection, as well as his stories of the life of civilians working at the base. The same can be said of Beatrice (Ramsden) Robertson of Kingston, who provided the pictures on page 49. John Yaney served as a fountain of information for the South Weymouth base as well.

Finally, we could not have finished the book without the help of Rockland's Paul Hartigan, who will forever be known to us as the guy who organized the airshows at South Weymouth. His collection, opened up to us at the last moment, blew us away.

John Galluzzo would like to thank Michelle Degni. Donald Cann would like to thank his daughters Emily and Jessica Cann and our "editor in chief," Janet Cann.

The stories of the life, death, heroism, and patriotism of the men and women who served and worked at the naval air facilities at Squantum and South Weymouth are being preserved by the formation of a museum on the grounds of the base. The authors are pledging a portion of the proceeds from the sale of this book to help support that cause, and they urge all who have photographs, memorabilia, or memories of the bases to make sure they find a proper permanent home. Although the bases may be gone, their stories can live forever. To donate items, or for more information, contact Bill Horsch at 781-878-3053.